641.35   Hughes, Meredith
HUG   Sayles.

Green power.

$26.60   11BT02749

DATE			

Cabbage
Broccoli
Artichokes
Spinach
Belgian Endive
Lettuce

Cabbage

PLANTS WE EAT

Broccoli
Artichokes
Spinach
Belgian Endive
Lettuce

Cabbage
Broccoli
Artichokes
Spinach
Belgian Endive
Lettuce

Cabbage
Broccoli

Green Power

Leaf & Flower Vegetables

Meredith Sayles Hughes

Lerner Publications Company/Minneapolis

Check out the author's website at www.foodmuseum.com

Lerner Publications Company
A division of Lerner Publishing Group
241 First Avenue North
Minneapolis, MN 55401 U.S.A.

Website address: www.lernerbooks.com

Designer: Edward Mack
Editors: Kari A. Cornell, Heather Hooper
Photo Researchers: Melanie Alvar and Cynthia Zemlicka

LIBRARY OF CONGRESS CATALOGING-IN-PUBLICATION DATA

Hughes, Meredith Sayles.
    Green power: leaf & flower vegetables / by Meredith Sayles Hughes.
      p.   cm. — (Plants we eat)
    Includes index.
    Summary: Describes the history, growing requirements, uses, and food value of various leafy green vegetables and vegetable flowers, including cabbage, broccoli, artichokes, spinach, Belgian endive, and lettuce. Includes recipes.
    ISBN 0-8225-2839-8 (lib. bdg. : alk. paper)
    1. Vegetables—Juvenile literature. 2. Greens, Edible—Juvenile literature. 3. Cookery (Vegetables)—Juvenile literature. [1. Vegetables. 2. Greens, Edible.] I. Title. II. Series: Hughes, Meredith Sayles. Plants We Eat
SB324.H844 2001
641.3'5—dc21   99–031716

Manufactured in the United States of America
1 2 3 4 5 6 – JR – 06 05 04 03 02 01

The glossary on page 77 gives definitions of words shown in **bold type** in the text.

Contents

Introduction

Plants make all life on our planet possible. They provide the oxygen we breathe and the food we eat. Think about a burger and fries. The meat comes from cattle, which eat plants. The fries are potatoes cooked in oil from soybeans, corn, or sunflowers. The burger bun is a wheat product. Ketchup is a mixture of tomatoes, herbs, and corn syrup or the sugar from sugarcane. How about some onions or pickle relish with your burger?

How Plants Make Food

By snatching sunlight, water, and carbon dioxide from the atmosphere and mixing them together—a complex process called **photosynthesis**—green plants create food energy. The raw food energy is called glucose, a simple form of sugar. From this storehouse of glucose, each plant produces fats, carbohydrates, and proteins—the elements that make up the bulk of the foods humans and animals eat.

Sunlight peeks through the branches of a plant-covered tree in a tropical rain forest, where all the elements exist for photosynthesis to take place.

First we eat, then we do everything else.

—M. F. K. Fisher

Plants offer more than just food. They provide the raw materials for making the clothes you're wearing and the paper in books, magazines, and newspapers. Much of what's in your home comes from plants—the furniture, the wallpaper, and even the glue that holds the paper on the wall. Eons ago plants created the gas and oil we put in our cars, buses, and airplanes. Plants even give us the gum we chew.

On the Move

Although we don't think of plants as beings on the move, they have always been pioneers. From their beginnings as algaelike creatures in the sea to their movement onto dry land about 400 million years ago, plants have colonized new territories. Alone on the barren rock of the earliest earth, plants slowly established an environment so rich with food, shelter, and oxygen that some forms of marine life took up residence on dry land. Helped along by birds who scattered seeds far and wide, plants later sped up their travels, moving to cover most of our planet.

Early in human history, when few people lived on the earth, gathering food was everyone's main activity. Small family groups were nomadic, venturing into areas that offered a source of water, shelter, and foods such as fruits, nuts, seeds, and small game animals. After they had eaten up the region's food sources, the family group moved on to another spot. Only when people noticed that food plants were renewable—that the berry bushes would bear fruit again and that grasses gave forth seeds year after year—did family groups begin to settle in any one area for more than a single season.

Organisms that behave like algae—small, rootless plants that live in water

It's a Fact!

The term *photosynthesis* comes from Greek words meaning "putting together with light." This chemical process, which takes place in a plant's leaves, is part of the natural cycle that balances the earth's store of carbon dioxide and oxygen.

Native Americans were the first peoples to plant crops in the Americas.

Domestication of plants probably began as an accident. Seeds from a wild plant eaten at dinner were tossed onto a trash pile. Later a plant grew there, was eaten, and its seeds were tossed onto the pile. The cycle continued on its own until someone noticed the pattern and repeated it deliberately. Agriculture radically changed human life. From relatively small plots of land, more people could be fed over time, and fewer people were required to hunt and gather food. Diets shifted from a broad range of wild foods to a more limited but more consistent menu built around one main crop such as wheat, corn, cassava, rice, or potatoes. With a stable food supply, the world's population increased and communities grew larger. People had more time on their hands, so they turned to refining their skills at making tools and shelter and to developing writing, pottery, and other crafts.

Plants We Eat

This series examines the wide range of plants people around the world have chosen to eat. You will discover where plants came from, how they were first grown, how they traveled from their original homes, and where they have become important and why. Along the way, each book looks at the impact of certain plants on society and discusses the ways in which these food plants are sown, harvested, processed, and sold. You will also discover that some plants are key characters in exciting high-tech stories. And there are plenty of opportunities to test recipes and to dig into other hands-on activities.

The series Plants We Eat divides food plants into a variety of informal categories. Some plants are prized for their seeds, others for their fruits, and some for their underground roots, tubers, or bulbs. Many plants offer leaves or stalks for good eating. Humans convert some plants into oils and others into beverages or flavorings.

People who frequently remind us to "eat our veggies" have in mind the green power plants we'll explore in this book—cabbage, broccoli, artichokes, spinach, Belgian endive, and lettuce. Cabbage, spinach, Belgian endive, and lettuce are leafy green vegetables that grow on a single stalk and have abundant leaves. Some of these plants, including cabbage, Belgian endive, and iceberg lettuce, grow heads—layers of leaves that are wrapped tightly around a central core. Other leafy greens, such as spinach, leaf lettuce, and some varieties of cabbage, grow in thick clusters of loose leaves.

Have you ever eaten a flower? If you've munched on crisp broccoli you have. Broccoli, cauliflower, and brussels sprouts, all members of the cabbage family, are vegetable flowers. Broccoli and cauliflower are fully formed flowers that grow from a single stem. Brussels sprouts are small flower buds that develop along the stalk. Artichokes aren't related to cabbage, but the spiky delicacy is also an immature flower head.

The plants featured in *Green Power* are not all members of the same vegetable family, but they do share common qualities. The green power plants contain powerful nutrients, more than most other food plants. Most of these plants are annuals, which means they must be grown directly from seed each year. Each green plant grows from a single sturdy stalk, but neither the stalk nor the underground root is important to nutrition. The leaves and the flowers have it all.

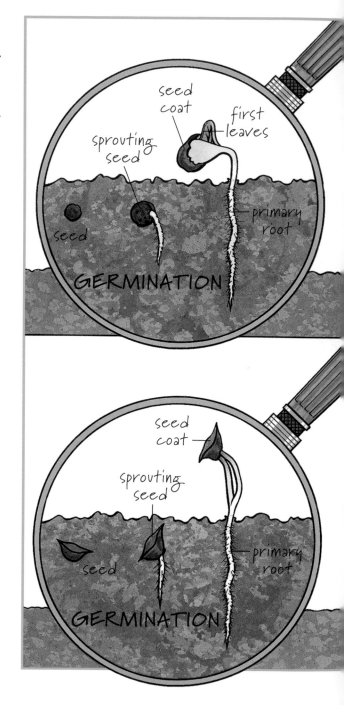

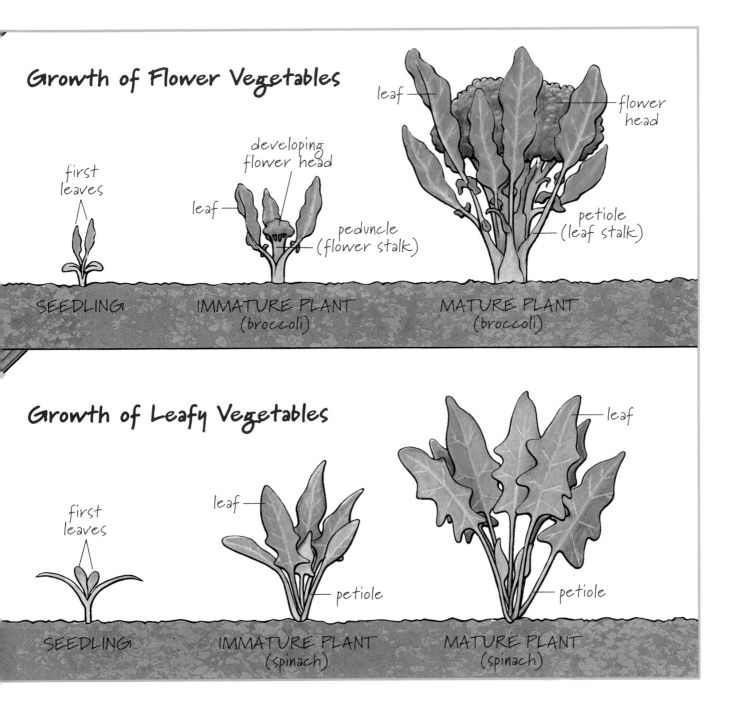

Growth of Flower Vegetables

first leaves

developing flower head

leaf

peduncle (flower stalk)

leaf

flower head

petiole (leaf stalk)

SEEDLING

IMMATURE PLANT (broccoli)

MATURE PLANT (broccoli)

Growth of Leafy Vegetables

first leaves

leaf

petiole

leaf

petiole

SEEDLING

IMMATURE PLANT (spinach)

MATURE PLANT (spinach)

Cabbage

[*Brassica oleracea capitata*]

The cabbage is one of the oldest edible plants around. The harvested core or head of the cabbage plant does look a bit like a human head. Large purple or green leaves, wrinkled or flat depending on the variety, cover the core in layers. The plant's resemblance to a large flower bud on a sturdy stalk prompted Japanese **botanists** in the 1960s to develop **ornamental,** not-for-eating cabbages in yellow, red, pink, and green varieties.

As an inexpensive vegetable that people tend to overcook, cabbage has earned a reputation for being ordinary and unsophisticated. Yet throughout history, people around the world have happily eaten pounds of it each year.

At a cabbage patch in Great Britain, cabbage is ripe and ready for harvest.

And he wore over all, as a screen from bad weather, a cloak of green cabbage leaves stitched all together.

—Edward Lear

Diogenes, a Greek philosopher (ca. 412–323 B.C.), is said to have spent most of his life in a tub, where he survived on cabbage alone.

Cool Heads

A cool-climate plant that loves humidity, wild cabbage probably first grew along the coasts of northern Europe more than ten thousand years ago. Historians believe that cabbages thrived in the north for generations. At some point, people or animals carried the seeds south and east to the Mediterranean coast. Birds and mammals played a big part in moving plants to new areas by eating plant materials and then leaving a few seeds behind in their droppings. Some seeds took root in fertile ground.

From Confucius's writings we know that the Chinese had cabbage, probably what eventually became known as Chinese cabbage, in the 400s B.C. For the Chou dynasty, Confucius wrote more than three hundred songs in which he mentioned forty-four different plants, including cabbage.

Before long, the Greeks discovered cabbage, but they didn't seem too elated about it. Eudemus, who studied under the philosopher Aristotle, may have been the first to write about cabbage. In a work dating from the 300s B.C., Eudemus described many plants, including three types of cabbage.

Keep Your Heads about You

Both the Greeks and the Romans felt that eating cabbage would prevent drunkenness. Although sources give no clues as to how they came to believe this, modern books on nutrition state that cabbage contains sulfur and chlorine—two minerals that clear the body of toxins. Could the Greeks and Romans have known this? Anything is possible. Roman gardeners developed many of the cabbage varieties popular in modern times, including giant heads weighing more than twenty pounds.

Cato (234–149 B.C.), a Roman statesman whose diet consisted mostly of cabbage, lived long and felt that cabbage was a cure for many ills, especially if eaten raw with vinegar. Much later, the Roman emperor Claudius (10 B.C.–A.D. 54), known for eating too much, was particularly fond of corned beef and cabbage.

Aside from Roman interest in the cabbage, we read little about the healthy heads until the early Middle Ages (1100s and 1200s). Botanists of the time argued about the merits of head-forming cabbage over leafy cabbage. From what we tend to eat in modern times, it's clear that heads won. Cabbage appears to have been a staple item, stored over the long winters and often served up stewed on its

Family Matters

To keep things straight in the huge families of plants and animals, scientists classify and name living things by grouping them according to shared features. These various characteristics become more noticeable in each of seven major categories. The categories are kingdom, division or phylum, class, order, family, genus, and species. Species share the most features in common, while members of a kingdom or division share far fewer traits. This system of scientific classification and naming is called taxonomy. Scientists refer to plants and animals by a two-part Latin or Greek term made up of the genus and the species name. The genus name comes first, followed by the species name. Look at the cabbage's taxonomic name on page 10. Can you figure out to what genus the cabbage belongs? And to what species?

THREE CABBAGE VARIETIES

SAVOY CABBAGE PLANT

WHITE CABBAGE PLANT

head

head

RED CABBAGE HEAD (cross section)

stem

On a huge farm near Saint Petersburg, Russia, two Russian women pick cabbages and load them onto a truck.

President Cabbage Head

For a long time, the Dutch and cabbages went hand in hand. Martin Van Buren, president of the United States from 1837 to 1841, was a native Dutch-speaker. Can you guess how he appeared in the political cartoons of the time? Cartoonists portrayed Van Buren as a cabbage head.

own or mixed with root crops, such as carrots, turnips, parsnips, and beets, in the soup pot. Because cabbage kept for long periods, it was often the only "green" still available to the majority of the population during the winter months. Its juice was sometimes used as a cough syrup and its leaves as a plaster or poultice on wounds.

Flemish weavers fleeing Spanish rule in Holland probably brought cabbages and cauliflower to England in the late 1550s. The English had probably already cultivated some types of cabbages, since the cabbage-loving Romans had once occupied the area.

The English word cabbage comes from an old French word caboche, meaning "head."

A Thai worker harvests a crop of Chinese cabbage grown on raised beds.

The earliest settlers on North America's eastern shores came from England, the Netherlands, Germany, and Scandinavia. They brought cabbage seeds with them. The Dutch established the colony (settlement) of New Netherland in about 1621 and grew cabbage along the Hudson River. In the capital city of New Amsterdam, modern New York City, citizens ate *spek ende kool*—pork with cabbage. At about the same time, their German neighbors in Pennsylvania and New Jersey made vats of pepper hash—pickled cabbage mixed with American peppers.

The medieval custom of using cabbage as medicine held on well into the 1700s. En route to explore Tahiti in 1769, the British doctor on Captain James Cook's ship, *Endeavor,* applied cabbage plasters to the wounds of forty crew members who had been injured in a horrific storm. Cook always carried abundant cabbages on board to prevent scurvy, the disease caused by lack of vitamin C. But he also favored canned sauerkraut. Before long the British navy established sauerkraut as an official food that was available on Royal Navy ships.

Off with Their Heads!

Cabbages are big business, associated with cool weather countries such as Germany, Poland, and Russia. But China leads the world in cultivation. In 1998 Chinese farmers grew 18,645,000 tons of cabbage!

In the United States, Wisconsin produces the most cabbage for processing. Most processed cabbage is used to make sauerkraut. Florida grows the freshest cabbage. Cabbage farmers in Florida are lucky. They have plenty of rich, black mucky soil close to lakes in the southern part of the state, where most of Florida's cabbage crop thrives. Growers have built canals to carry lake water into the fields. To keep moisture levels fairly steady, growers use pumps to adjust water levels.

Under normal conditions, growers who begin planting in November can harvest two cabbage crops—one that same winter and the other the next spring. Growers place the cabbage seeds directly into the soil using mechanical seed drills. Machines apply chemical herbicides to keep the fields weed free until the heads are large enough for cutting. After fifty-five to seventy days, the cabbage is mature. Workers still cut cabbage by hand. Several workers walk behind a truck that drags a long conveyor boom, a beam that extends from the truck and supports a conveyor belt. They bend over and use machetes (long knives) to lop off the cabbage heads. Workers toss the harvested heads onto the conveyor belt. Some workers sit on the boom, where they trim extra leaves from the heads and then pack the cabbages into boxes.

A boom is used to guide cargo.

It's important to refrigerate or cool the harvested cabbage immediately to keep it fresh. Growers may choose to cool the boxes in the field by having workers spray the cabbages with cool water. Other growers prefer to have workers take the boxes directly to a nearby cooling shed, where the product sits until it's shipped to the customer. Some packing companies sell to food brokers, who sell the produce to grocery store chains. Others sell directly to stores.

After harvest, growers use a machine to disk, or cut up, the excess leaves, stalks, and roots left in the ground and to plow the pieces back into the soil. The decaying plant material enriches the soil. Some farmers also plant **cover crops** such as clover or vetch in the off-season, when cabbage isn't growing. Then they plow under the plants to fertilize the soil before planting more cabbage the next fall. Other growers may use chemicals to feed the soil.

Workers toss harvested heads on the conveyor belt.

Wild and Fruitful

The wild cabbage, native to coastal areas in Great Britain and along the Mediterranean and Adriatic Seas, is a parent plant to many unique veggies such as cabbages, cauliflower, collards, broccoli, brussels sprouts, kale, and kohlrabi. Food historian Waverley Root has pointed out that farmers developed these vegetables over the centuries by enhancing certain traits that already existed in the original cabbage plant. First, farmers cultivated wild cabbage, a plant with curled leaves, to grow into "heads." Then growers encouraged cabbage to flower to create broccoli and cauliflower. Growers developed brussels sprouts by coaxing cabbage buds, which appear where the plant's leaves meet the stem, to form into mini cabbages. From one plant, growers created several spin-off plants.

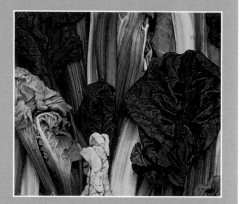

Let Them Eat Heads!

People eat cabbage cold or hot, shredded or whole. Cooks around the world wrap cabbage leaves around assorted fillings and cook them or stuff whole cabbages and steam them. The French gently cook their *chou farci,* whole cabbage filled with sausage and left-over meat, in beef consommé (a thin soup) for three hours. Scandinavians do just about the same, adding rice toward the end. Serbs stuff cabbage leaves with pork and beef and

The cabbage family has many members, including broccoli, cauliflower, and brussels sprouts.

To Your Health!

The cruciferae, any plant that belongs to the mustard family, are health-giving plants. Rich in vitamin C and fiber, they contain antioxidants, which attack cancer-causing elements in the body. The cole crops—including broccoli, kale, brussels sprouts, cabbage, cauliflower, and kohlrabi—are especially effective against colon cancer.

Cabbage juice prevents ulcers, strengthening the stomach lining to withstand stomach acids.

This Alaskan woman proudly displays her sixty-pound cabbage.

then bake them with spareribs and bacon. Italian cooks make *cavolo imbottito,* cabbage leaves stuffed with pork, beef, Romano cheese, and spinach. Algerians make cabbage rolls stuffed with lamb and cinnamon. In Portugal, cooks season cabbage rolls with black olives and tomato sauce.

Shredded cabbage is popular around the world, too. The Chinese stir-fry both head cabbage and Chinese cabbage with other assorted vegetables. Indonesians lightly cook their *gado gado*—steamed cabbage, broccoli, carrots, yams, and green beans mixed with rice—and serve it with a peanut sauce. *Rode kool* is a simple Dutch dish of cooked red cabbage sweetened with brown sugar and apples. A French version uses chestnuts, while Bavarians add bacon. Malaysians cook shredded cabbage with coconut milk, chilies, and fish paste. A classic Irish dish is *colcannon*—cooked chopped cabbage mixed with mashed potatoes, scallions or leeks, and parsley.

Hearty cabbage soups bring warmth in winter. The Basques, a people from the mountainous region along the French and Spanish border, specialize in a white bean and cabbage soup. Belgians traditionally prepare *hutsepot* at Christmas—soup that blends pork sausage, turnips, and cabbage. Estonians make their *hapukapsa supp* with sauerkraut, oats, and bacon.

In Taiwan, greengrocers travel from town to town to sell many leafy greens, including cabbage.

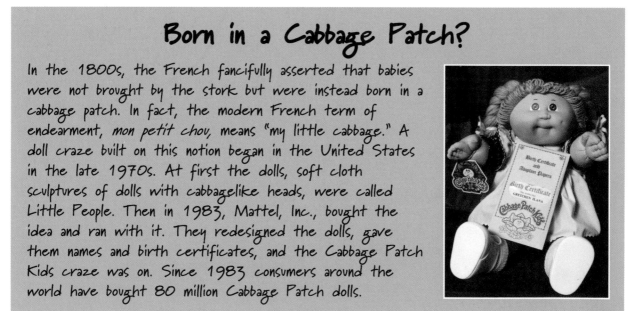

Born in a Cabbage Patch?

In the 1800s, the French fancifully asserted that babies were not brought by the stork but were instead born in a cabbage patch. In fact, the modern French term of endearment, *mon petit chou*, means "my little cabbage." A doll craze built on this notion began in the United States in the late 1970s. At first the dolls, soft cloth sculptures of dolls with cabbagelike heads, were called Little People. Then in 1983, Mattel, Inc., bought the idea and ran with it. They redesigned the dolls, gave them names and birth certificates, and the Cabbage Patch Kids craze was on. Since 1983 consumers around the world have bought 80 million Cabbage Patch dolls.

Dig In!

REUBEN SANDWICH
(MAKES 2 SERVINGS)

A great grilled sandwich that's easy to make, the Reuben was probably created in 1914 by Arnold Reuben, owner of a famous New York delicatessen. This recipe makes two sandwiches.

4 slices sourdough, rye, or pumpernickel bread
4 slices Swiss cheese
¼ pound thinly sliced corned beef, or in a pinch, ham
4 ounces (half a can) of sauerkraut, drained
4 tablespoons of Thousand Island dressing
butter

Butter one side of each bread slice. Place two slices buttered-side-down in large skillet over medium heat. Place one slice of cheese on top of each piece of bread. Then add meat and sauerkraut and top each sandwich with another piece of cheese. Spread two tablespoons of dressing over the cheese. Place the remaining two bread slices butter-side-up on top of each sandwich. Continue cooking until bottom slice browns. Use a pancake turner to gently lift each sandwich to check for doneness. With the help of an adult, carefully flip each sandwich with the pancake turner and grill the other side. Eat hot. This sandwich also tastes wonderful without the meat.

Broccoli

[*Brassica oleracea italica*]

Some people, including former U.S. president George Bush, simply do not like broccoli, and they let you know it. But a recent poll of middle school students revealed that they not only like broccoli—stir-fried, please—but also prefer it to many other vegetables.

A fairly low-growing plant, most varieties of broccoli are a rich green, although some are purple. The broccoli plant has long, thin, crinkly, silvery green leaves. It sends up stalks that resemble a child's drawing of a tree—with a straight light green trunk and a bushy green top. The stalks and flowering heads are what we eat.

The word *broccoli* comes from an Italian word *braccio*, which means "arm."

The local groceries are all out of broccoli, loccoli.

—Roy Blount Jr.

Branching Out

Very little was written about broccoli's early days. One food historian has suggested that because the ancient Romans wrote nothing about the creation of or arrival of broccoli it seems likely it already grew in Etruria (modern Tuscany) when the Romans conquered the region in about 200 B.C. The famous Roman cook and creator of recipes, Apicius, was said to have prepared it well. A typical Roman evening meal might well have included some meat or fish and a grain pancake with onions and broccoli. Emperor Tiberius, who ruled from 14 B.C. to A.D. 37, had a son named Drusus who loved broccoli. Evidently the emperor had to warn Drusus he was eating too much broccoli.

The historical record on broccoli leaps from the Roman Empire to France in the 1500s. Catherine de Médicis was a fourteen-year-old Italian aristocrat who traveled from Florence, Italy, to Paris in 1533 to marry the future king of France, Henry II, also fourteen. Many servants and chefs accompanied Catherine. Catherine's chefs brought with them Italy's finest fresh vegetables, seeds with which to grow them, and recipes. Among these fresh vegetables was broccoli. (And green beans, artichokes, cabbage, and more. . . .)

By 1721 English gardeners were growing broccoli. Known at that time as "Italian asparagus," broccoli became popular throughout Great Britain primarily because the French—whom the British looked to for new trends in food—liked it. And you can bet that if the French, British, and Italians liked it, U.S. president—and experimental gardener—Thomas Jefferson was bound to have it in one of his gardens. Sure enough, on May 27, 1767, Jefferson noted in his gar-

It's a Fact!

Since the 1980s, the farming community of Greenfield, California, has hosted the Broccoli Festival over Labor Day weekend each September. Vendors serve up broccoli ice cream, deep-fried broccoli—served one year with rattlesnake—broccoli burgers, and baked potatoes topped with a broccoli cheese sauce. Adults can even wash it all down with broccoli margaritas.

to the Italian American community. But it was broccoli that put them on the map. The D'Arrigos' father shipped broccoli seeds from Italy. The brothers became the first commercial growers in the western United States to successfully raise and ship box loads of the plant. Soon they created a distinctive brand name for their broccoli—Andy Boy. A photo of Stefano's two-year-old son, Andrew, decorated the label. The D'Arrigos formed the first fresh produce company in the United States to use a brand name in its advertising.

Cutting Down Trees

Broccoli is big business in Italy, France, and China. In the United States, Arizona and Oregon grow broccoli, but the bulk of U.S. broccoli is grown in California. Aroostook County, Maine, known more for its potato crop, is becoming a major supplier of broccoli as well. Americans eat about four pounds of broccoli per person each year. The United States exports $85 million worth of broccoli, primarily to Canada and Japan.

California growers plant broccoli seeds directly into fine-textured soil twice each year—once in February and again in September. Tractor-pulled machines called drills or vacuum planters place the seeds three to five inches apart in rows anywhere from fourteen to thirty inches apart. In California it takes ten to fourteen inches of irrigated

Some say broccoli grew in ancient Tuscany, a region in northern Italy that was once called Etruria. This Etruscan wall painting of a warrior still exists from ancient times.

den book that he planted broccoli, lettuce, radishes, and cauliflower.

Italian Americans were more than likely growing broccoli in home gardens from the early 1900s on. The crop was not widely known in the United States until some years later, however.

Two brothers, Stefano and Andrea D'Arrigo from Messina, Italy, arrived in the United States in the early 1900s. They worked at various jobs before learning the grape juice business. In 1922 they started their own produce company in San Jose, California. Their specialty was vegetables and fruits familiar

Workers pack harvested broccoli into Andy Boy boxes.

water and about eighty to one hundred and twenty days for thirty thousand plants to grow to maturity per acre.

At harvesttime a worker slowly drives a tractor through the rows, pulling a long boom or platform. Each field is harvested two to three times per growing cycle. Fifteen to twenty people walk behind, cutting off the broccoli stalks with triangularly shaped knives and tossing the stalks into bins on the boom. Eight other workers stay on the boom, sorting and packing the stalks into boxes, which are often made of cardboard covered in wax. A separate tractor pulls a trailer right alongside the boom. Farm laborers take the filled boxes off the boom and stack them on the trailer. Broccoli is highly perishable, so when the trailer is full, the tractor driver rushes the packed boxes to nearby cooling sheds.

Go Gai Lin!

Chinese kale, also known as Chinese broccoli, is called *gai lin* or sometimes *gai lohn* in Chinese. In March and April, the greengrocers of San Francisco's Chinatown pack their counters with mounds of *gai lin*. A leafy green, it is usually sold with its central skinny stalks topped with half-opened yellow blossoms. In nearby restaurants, cooks chop bushels of *gai lin* into pieces about four inches long and stir-fry them swiftly in woks, pans shaped like bowls. *Gai lin* is served hot and crisp, often with oyster sauce or sesame oil.

Broccoli fields are planted in rows for better irrigation.

At the cooling sheds, a worker uses an injection nozzle to fill each box with liquid ice, a combination of crushed ice and water. The ice cools the crop's temperature to about 35°F in a matter of five minutes. Workers then load boxes onto trucks for delivery, either to food brokers or directly to supermarket warehouses. To keep the broccoli fresh, growers ship the boxes immediately so that the produce is in stores within three to five days of picking.

Some growers send broccoli to frozen food specialists to be rapidly processed into a frozen product. At frozen food processing plants, workers quickly **blanch,** or boil the vegetable briefly, and then freeze and package it. Other workers may chop a load of broccoli into bite-sized florets and pack them in large clear plastic bags. At some plants, processors make broccoli slaw—a new twist on the coleslaw concept. Broccoli slaw is made with shredded broccoli stems instead of shredded cabbage leaves.

Workers gather handpicked broccoli in special backpacks *(below).* When they are full, workers quickly load the broccoli onto a boom *(above)* so it can be rushed to a cooling shed.

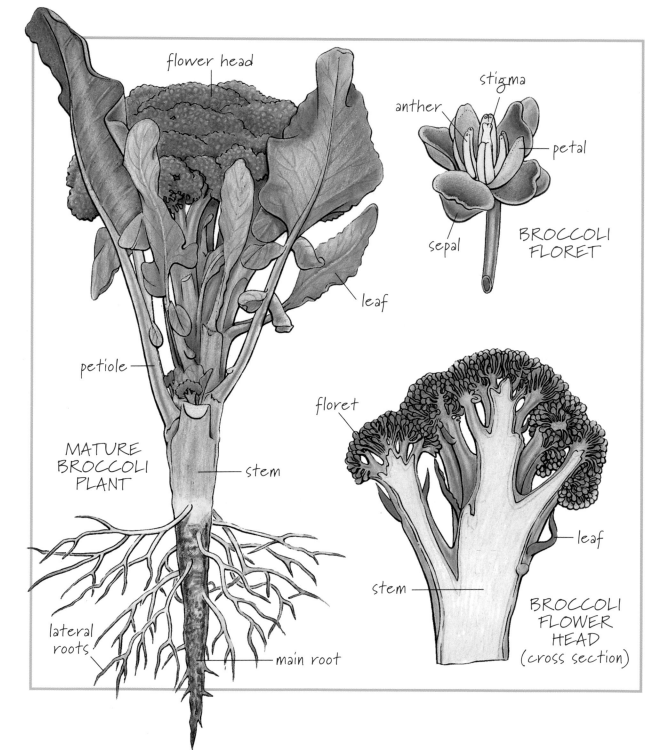

flower head

stigma

anther

petal

sepal

BROCCOLI
FLORET

leaf

petiole

floret

MATURE
BROCCOLI
PLANT

stem

leaf

stem

lateral
roots

main root

BROCCOLI
FLOWER
HEAD
(cross section)

A broccoli harvesting machine moves through a field while workers check the broccoli for ripeness.

Tiny Broccoli?

It's broccolini—a hybrid (mix) of broccoli and Chinese kale. With long and slender stalks topped with small buds, broccolini has a sweeter taste than broccoli.

Broccolini was an eight-year project for the California seed company Sakata Seed. Sakata originally wanted to create a vegetable similar to broccoli that would thrive during the hottest months of the year. In 1989 the company decided to start breeding broccoli with Chinese kale, a heat-tolerant green. By **hand-pollinating** hundreds of generations of plants, Sakata achieved its first hybrid after seven years. The company then shipped out the new seed to growers around the world. Ironically, the new veggie has not proven to be a hot-weather replacement for broccoli after all. Nonetheless, broccolini appears to have slowly gained fans in the restaurant and gourmet grocery business since its introduction in 1997.

Broccoli is a popular and healthy ingredient in stir-fry dishes.

Steam 'til Green

Broccoli can be steamed, briefly boiled, stir-fried, or baked in casseroles. Overcooking robs broccoli of its bright green color and creates a strong cabbagelike taste. Thai cooks lightly sauté broccoli with garlic, chilies, and spring or green onions. *Broccoli alla Romana* is Rome's specialty. Italian cooks simmer broccoli in white wine and garlic. In Italy's far south, cooks sauté broccoli and garlic in olive oil. When served with pasta and topped with grated cheese, the mix makes a tasty dish. North Americans sometimes make broccoli, egg, and cheese casseroles or dip fresh broccoli tops into a sour cream dill sauce.

To Your Health!

Broccoli is one of the powerhouses of the vegetable family. It contains a chemical compound called sulforaphane, which slows cancer development. The compound advances on carcinogens (cancer-causing agents) before they begin to eat away at the body. Broccoli sprouts contain 100 percent more sulforaphane than mature broccoli. Broccoli is also loaded with vitamins C and A and iron. Vitamins C and A help the body fight illness. Iron keeps oxygen levels constant in red blood cells.

Dig In!

GARLIC BROCCOLI WITH RIGATONI PASTA (4 SERVINGS)

Broccoli, cooked lightly and topped with fresh Parmesan cheese, mixes well with any pasta. Rigatoni are short, fat, ridged, macaronilike pieces of pasta.

1   12-ounce box rigatoni pasta
2   tablespoons olive oil
15–20 medium-sized broccoli stalks, or florets
4   garlic cloves, crushed with the flat of a knife, peeled, and then chopped
2   tablespoons water
¾ cup freshly grated Parmesan cheese

Fill a large kettle with hot water and bring to a boil. Add a tablespoon of olive oil to the water before adding pasta. Cook rigatoni for about 15 minutes, drain and return to the pot. Add another tablespoon of oil to the pot and shake to coat pasta with it. Meanwhile, wash, drain, and trim broccoli stalks and heads into bite-sized pieces. In a heavy skillet, cook crushed garlic and broccoli at high heat for about one minute, moving the food around gently with a wooden spoon. Then add water, cover the pan, and reduce heat to medium. Check stalks for tenderness after another minute. When broccoli can be pierced easily with a fork, remove from heat at once. Place pasta in individual large soup plates, add broccoli and garlic topping, and then sprinkle with freshly grated Parmesan cheese. A dash of olive oil on top of the cheese is a tasty extra touch. Serve piping hot.

Artichokes

[*Cynara scolymus*]

The artichoke plant looks like a thistle all the way down to the plant's prickly leaves. But it is actually a member of the daisy family. A plant that originated along the Mediterranean Sea, the artichoke produces large flowers. People love to eat the flower's tender petals and its soft but hard-to-get-to center.

Sometimes known as globe artichoke because of the shape of the bud (a small swelling on a plant stem from which a flower or leaf will grow), the artichoke plant grows three to five feet tall. Long gray-green leaves surround the bottom of tall stalks tipped with artichoke buds. If allowed to bloom, the artichoke flower opens in purple-pink glory.

Artichokes are actually flower buds.

I seek in anonymity's cloister, Not him who ate the first raw oyster, But one who, braving spikes and prickles, The spine that stabs, the leaf that tickles, With infinite patience and fortitude, Unveiled the artichoke as food.

—Ogden Nash

A stone carving depicts a Roman butcher in his shop. Wealthy Romans enjoyed artichokes as a part of their feasts.

A Spiky Discovery

One of many food mysteries is how people first fought their way past the spines and barbs of this plant to eat any part of it. But eat it they did. Artichokes have become increasingly popular, a fact that explains why this plant has survived as a cultivated crop.

No one really knows when growers first cultivated artichokes, but they more than likely grew them in Sicily, the island just off the southern tip of Italy. The Greeks occupied Sicily in the 700s B.C., and travelers from the island probably brought artichokes back to Greece with them. Scholars believe both the Greeks and the Romans had artichokes. But no one is absolutely certain whether the earliest writers were actually referring to the artichoke or to the cardoon, another thistle-like plant. Growers in Greece and in the Roman Empire's Mediterranean region grew both plants.

In A.D. 77, Pliny, a Roman naturalist, called the artichoke "one of the earth's monstrosities." But he and his colleagues continued to enjoy eating them. Wealthy Romans enjoyed artichokes prepared in honey and vinegar and seasoned with cumin, a spice originally from Egypt. They stored the mixture in jars so that the treat would be available year-round.

The Roman Empire eventually expanded to include Gaul (modern-day France), Egypt, and other areas in North Africa. The Roman newcomers brought artichoke plants with them. A later North African Arab people known as the Moors cultivated artichokes in North Africa as the Roman Empire declined. Another group of North African Arabs, the Saracens, cultivated artichokes in Sicily at about the same time.

Saved by the Moors

After the fall of Rome in A.D. 476, we lose track of artichokes until about A.D. 717, when the Moors headed north to occupy the southern part of the Iberian Peninsula. The North African newcomers carried many plants with them across the Mediterranean Sea to Spain and Portugal. By A.D. 800, the

Moors had begun to cultivate artichokes in the area of Granada, Spain.

By 1500 it's probable that monks in Tuscany had improved the artichoke in their walled gardens. The monks transformed the artichoke from a marginal food into the much more edible plant we recognize in modern times.

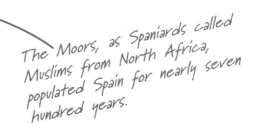

The Moors, as Spaniards called Muslims from North Africa, populated Spain for nearly seven hundred years.

The artichoke pops up in the record again in 1533 with Catherine de Médicis. Remember her? Artichokes were young Catherine's favorite food. This particular craving scandalized elders in the court. They considered the artichoke an "improper" food, probably because diners had to peel the vegetable petal by petal, which can be a very messy job.

But if royals favored it, it was fashionable. If it was fashionable, then wealthy people ordered their private gardeners to grow it. Soon the artichoke was popular in France.

English farmers tried growing artichokes in 1548. But they didn't make a huge splash there or anywhere else too far from the Mediterranean region. In the 1600s, Spanish settlers brought the artichoke to California, where it grew in small individual gardens. Commercial growers did not introduce the plant there until the 1920s.

The young Catherine de Médicis, a future queen, scandalized the elders of the court with her love of artichokes.

To Your Health!

Artichokes are rich in iodine, an element most frequently found in edible seaweeds. Iodine is important to the healthy functioning of the thyroid, a gland in the neck that regulates the body's use of energy. Chokes also contain substantial amounts of potassium, which regulates the amount of water in the body and helps build muscle. Best of all, artichokes are extremely low in calories—before you dip them in that lovely rich sauce, that is.

Castroville, California

Choking on Cash

In 1922 Andrew Molera, a landowner in the Salinas Valley of Monterey County in northern California, decided to lease land once used to grow sugar beets to farmers willing to grow artichokes. His reasons were economic—artichokes were already fetching high prices from grocery store suppliers. Artichoke farmers could pay Molera three times what the beet growers could for the same land. By 1929 artichokes were the third-largest produce crop in the valley.

In modern times, artichokes are still big business in Monterey County. The tiny town of Castroville sits in the center of vast acres of artichoke fields. It is home to the country's only artichoke-processing plant and to the annual Artichoke Festival, which began in 1959. The festival features races, music, a parade, and of course, lots of fresh artichokes—steamed, fried, sautéed, and made into soup.

flower bud

flower (open)

flower bud

leaf

MATURE ARTICHOKE STEMS

stem

scale

peduncle

ARTICHOKE FLOWER BUD (exterior view)

scale

heart

ARTICHOKE FLOWER BUD (cross section)

The harvested artichokes are placed in wooden bins for transport *(left)*. Harvested artichokes are collected in backpacks *(right)*.

Stumped? Try Seeds!

In modern times, Italy grows the most artichokes in the world, averaging about 560,000 tons each year. In the United States, all commercially grown artichokes come from California. Farmers grow 45,000 tons each year—about 80 percent of them in Monterey County.

Traditionally, artichoke growers in California have raised the green globe artichoke using cuttings, called stumps, from the root of mature plants. Stumps can be as large as one foot tall and one foot wide. Each new plant grows from a stump. Between March and May, growers plant the stumps by hand in large holes, arranged in tidy rows. A **perennial,** which means it grows back each year, a commercial artichoke plant produces for five to ten years.

Growers increasingly seek reliable **hybrid** seed from which to raise artichokes. Why? For one, starting from seed is easier and involves less hand labor. Workers can plant the seeds in rows using a tractor. Once fully harvested, growers can plow under plants grown from seed to allow the field to rest or to plant it with a soil-nourishing cover crop. Perennial plants left to grow in the same ground for years are easy prey for insects, plant diseases, mice, or even floods.

Researchers estimate that between 85 to 95 percent of all California artichokes grow from stumps, and the rest grow from seed. In time, seed planting will probably replace the stump method, once farmers are satisfied that they are getting quality buds from plants grown from seed.

Cut Flowers

It takes approximately 125 days for artichoke buds to mature. Globe artichokes are hand tended by experienced farmworkers. As the artichokes ripen, workers use hoes to keep the base of the plants free from weeds. This allows the roots of the plants to breathe and ensures that fertilizers feed the crop and not the weeds.

March through May is the artichoke's peak harvest season. Workers pick the flow-

Artichokes, members of the daisy family, have petals, leaves, and stems.

The artichoke will blossom into a pink flower if it is allowed to ripen completely.

ers when their petals—usually called scales, each tipped with tiny spikes—are tightly closed. When buds begin to appear, workers walk through the field cutting them from the plants with a knife. They toss the artichokes into large collapsible backpacks. When the packs are full, workers empty them into wooden bins at the end of the rows. Since the plants develop buds at different times, workers walk the same field once a week during peak season. Workers first cut the largest chokes, which grow on the plant's central stem. Medium-sized buds, which are usually hidden under the leaves, are next. Medium-sized artichokes and even the smallest buds (known as "baby artichokes") are fully

Vacuum processing is a quick way to cool vegetables. The vegetables are placed in a tightly sealed room from which all of the air has been removed.

A grocer checks the quality of her artichokes. The best artichokes are those with tightly formed buds.

mature. When workers have harvested each plant, they use machetes or knives to cut the stems back several inches below the surface of the ground. This encourages new, strong growth. In October there is another, smaller harvest, when workers again cut buds from the same plants.

Growers must cool all fresh produce immediately after picking. Sometimes workers pack artichokes into cartons right in the field. Truck drivers then haul the cartons to a cooling shed, equipped with a huge boxlike machine known as the Hydro-Vac. Workers place the cartons inside the machine, which uses a vacuum process to lower the temperature of the produce to 34°F in about twenty minutes. Laborers remove the cooled cartons and stack them on small forklifts. Forklift operators move the cartons to a refrigerated loading dock area where the artichokes await delivery to distributors.

Other growers may bin and truck their artichokes directly to packing sheds. Workers there dump the artichokes from the bins into vats of room temperature water. (Cold water could negatively alter the artichokes' color.) From the vats, the chokes then float down water chutes (slides) into the sorting line, where workers pull out any imperfect artichokes. Workers sometimes chop up the rejected chokes, called "culls," and later plow them back into the ground as fertilizer or feed them to cattle. A local source claims the

Artichokes are an agreeable addition to a meal.

cows produce the sweetest milk around. From the sorting line, the vegetables move onto rubber rollers, where workers sort and pack the chokes into cartons.

Grocers sell most California artichokes fresh. But U.S. growers send many of the smaller chokes, known as "canners," to the country's only artichoke-processing plant, in Castroville, California. There workers dump the artichokes from wooden bins onto conveyor belts. From that point, the process is automated. Machines lop off the tops and bottoms of the chokes and remove their outer leaves. Then another cutting device slices the buds in half or in quarters. Workers stand by as machines blanch the processed vegetables by cooking them briefly in hot water and then stuff the artichokes into glass jars. Another machine pumps olive oil, garlic, herbs, and vinegar into the jars before sealing them. The jars move along on the belt and into the cooker, where they are covered in boiling water for a few minutes until sterilization is complete.

It's a Fact!

Artichoke hearts, those tender bits drenched in garlic and olive oil, are not really hearts. They're not even centers or cores. They are small artichokes that have been trimmed, cut in half, and soaked in olive oil.

Go for a Dip

Italian cooks probably have more artichoke recipes than anyone else. Diners eat chokes dipped in batter and deep fried; sautéed in garlic, lemon juice, and olive oil; mixed into omelets; as a topping on pizza; or as an elegant ingredient in salad. To prepare a favorite dish called *carciofi al tegame alla Romana*, cooks boil the artichokes, stuff them with mint, garlic, and vinegar-flavored breadcrumbs, and then bake them.

Greeks also enjoy this Mediterranean vegetable. To make *pilafi me anginares*, cooks simmer small artichokes in chicken broth and mix them with lemon and herb-flavored rice. French cooks like to stuff crêpes with small artichokes and ham. Sometimes the French take the soft meaty bottom, or fond, of the vegetable and stuff it with barely cooked new peas. North Americans tend to boil artichokes. They dip leaf after leaf into either garlic-flavored mayonnaise, hollandaise sauce, or melted butter. Californians also deep-fry them Italian style. In Louisiana over the Christmas season, cooks prepare stuffed whole artichokes using a recipe from Sicily. They use only the tenderest one-third of the artichoke in this dish. Cooks remove the toughest leaves, stuff the chokes with grated cheese, seasoned breadcrumbs, shallots, onions, and celery, and then simmer them in broth.

To Europeans and Asians, artichokes are more than just a tasty food. Many take

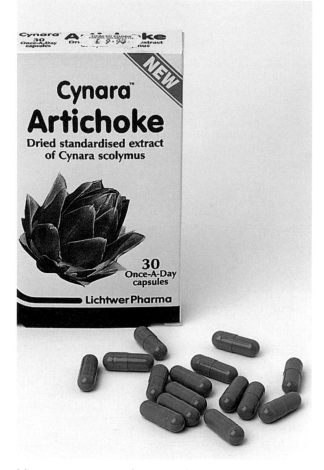

Many countries produce supplemental medicine tablets made from ground artichokes. These tablets are said to aid digestion.

tablets made from dried, ground artichokes to maintain healthy liver function and to lower blood cholesterol levels. Swiss, German, and Vietnamese companies have asked California artichoke growers to grow plants to be made into supplemental tablets.

Dig In!

BOILED ARTICHOKES
(4 SERVINGS)

Here's one of the simplest ways to prepare artichokes—just trim them, boil them up fast, and dip their leaves in a tasty sauce.

4 large fresh artichokes
3 lemon slices
mayonnaise or melted butter

Wash the artichokes, cut off the stems, and peel off the small lower leaves. Trim all the other spiky leaf points with scissors. Meanwhile, bring a large pot of water to a rolling boil. Toss lemon slices in the water to help maintain the fresh green color of the artichokes.

   Place the chokes carefully in the boiling water, letting them bounce around for about forty minutes at a full boil. Use a slotted spoon to remove one artichoke to check it for doneness. If the outer leaves come off easily and the bottom can be easily pierced with a fork, the artichokes are ready.

   Use the slotted spoon to remove the other artichokes from the water and place them on a plate to cool. Eat the artichokes by pulling off the leaves and dipping them in mayonnaise or melted butter. Use your teeth to pull the "meat" from the inside of each leaf. Once you've eaten most of the leaves, you'll reach the furry center of the choke, the part that really looks like a flower. Ask an adult to cut the furry part away from the fond carefully. The fond is the prized final bite of the vegetable—it's soft and very tasty and should come out easily with a spoon.

Spinach
[*Spinacia oleracea*]

Spinach is a dark green leafy plant with spade-shaped leaves—sometimes crinkly, sometimes flat—on slender stems. It grows anywhere from seven to twelve inches tall. The most famous spinach lover in the world is probably Popeye the Sailor Man, a cartoon character. Popeye first burst on the scene in 1929 in a strip developed by Elzie C. Segar. His affection for spinach out of the can sparked a 33 percent jump in spinach consumption in the United States during the early 1930s, the heart of the Great Depression. Grateful spinach growers of Crystal City, Texas, built a statue to the healthy sailor in 1937. Before long, other Popeye statues turned up—one in the cartoonist's hometown of Chester, Illinois, and another in front of the Alma, Arkansas, Chamber of Commerce building. Alma, a self-proclaimed Spinach Capital of the World, is the home base of Popeye Brand Spinach.

The English, French, Italian, and Spanish words for spinach all derive from the Arab word *isfanakh*.

I'm strong to the finish, cause I eats me spinach....

—Popeye the Sailor Man

This twelfth-century paper fragment shows the Fatimids—a political and religious dynasty that ruled parts of the Middle East and Africa—leaving their fortress to attack the Crusaders. Some say the Crusaders brought spinach to Europe.

The cartoonist himself was an avid spinach eater. Some suggest that Segar's doctor turned him on to spinach when the cartoonist was sick as a child. For years, spinach growers regularly sent Segar free cases. He finally had to ask them to stop the flow.

The Green Scene

Spinach probably originated in Persia (modern-day Iran), but when or in what form remains a mystery. Modern spinach may be related to a wild spinach that still grows in Iran.

While reports of spinach surfaced here and there in the historical record, most experts believe ancient gardeners were referring to some other leafy plant. In earlier times, English-speakers used the word *spinach* generically, as a term for all leafy greens.

An early agricultural record in China reports that the king of nearby Nepal gave spinach as a gift to the Chinese court in A.D. 647. The Saracens may have obtained the plant from Persian traders and carried it to the island of Sicily in the 800s. It's also possible that the Crusaders—Christian soldiers who fought Muslims for control of the Holy Land (Jerusalem) from about 1096 to 1270—brought spinach to Europe two centuries later, along with other foodstuffs they had gathered up in the Middle East.

By the 1200s, Italian farmers routinely planted spinach in the fall and harvested it in time for Lent—a period of fasting during the spring weeks before Easter. This fall-spring pattern agrees with spinach, since the vegetable thrives in cool weather. In hotter months, spinach plants begin to produce bitter-tasting leaves and go to seed. By 1351 the French were planting and eating spinach according to the Italian growing schedule. It's likely that monks, who tended spinach in their gardens, were the first to grow spinach in France. Spinach was one of the foods that French monks were allowed to eat on fast days. Fasting, which originally meant not eating at all, came to describe the voluntary giving up of certain foods for religious reasons. Monks usually gave up meat. Spinach, packed with iron, protein, and fiber, was a healthy alternative. The English were knee-deep in spinach from at least 1568, growing as much as the French. The English ate spinach, an early spring green, as a Lenten food, too.

Popeye and Olive Oil

Popeye Brand Spinach is the number-two selling brand in the United States. Popeye's bulging biceps also promote bags of fresh spinach. Did Popeye's creator, Elzie C. Segar, choose the name Olive Oyl for Popeye's scrawny girlfriend because olive oil and spinach go so well together?

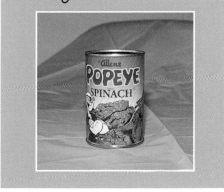

A Tunisian woman inspects a basket of spinach at a vegetable market in Nefta, Tunisia.

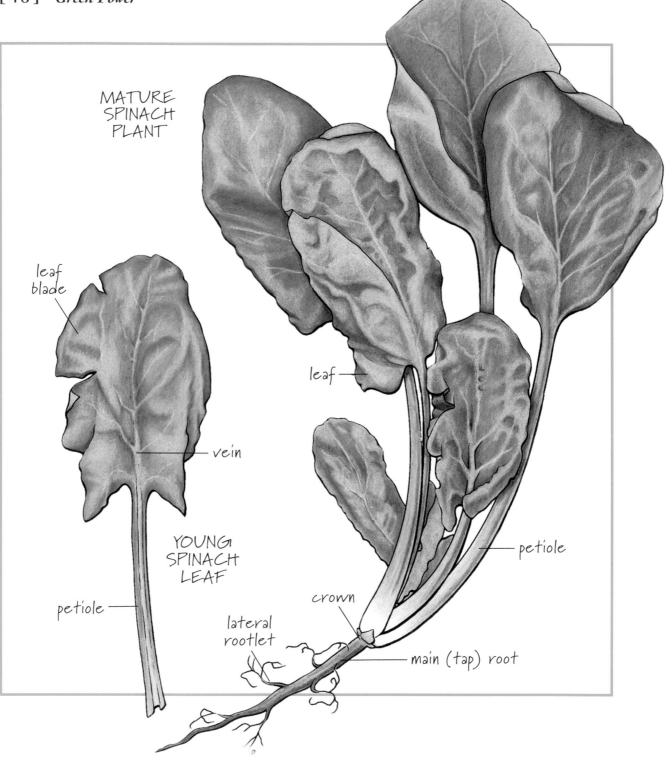

MATURE
SPINACH
PLANT

leaf
blade

vein

YOUNG
SPINACH
LEAF

petiole

leaf

petiole

crown

lateral
rootlet

main (tap) root

While there are different ways to plant spinach, most farmers choose to plant spinach in rows. The spinach in this field is fully mature.

It's a Fact!

African Spinach

In many parts of Africa, different leafy green plants may be prepared like spinach and may even be called spinach. For example, the leaves of the taro plant are known as spinach in Ghana and in other parts of West Africa. In Egypt cooks use a plant known as *molohia* interchangeably with spinach. People in Tanzania call *mchicha* (a locally grown green) spinach, too.

Spanish and other European immigrants may have carried spinach seed to the Americas in the 1600s. Immigrants often brought seeds and even young plants from their homelands. Historians have no specific record of when the first spinach seed sprouted in North America. On March 23, 1774, the future president of the United States and lifelong gardener, Thomas Jefferson, noted the planting of *spinaci*, Italian for "spinach," in his kitchen garden. This plot near the kitchen supplied fresh produce to the household.

For generations, American gardeners probably grew spinach on small plots and sold the vegetable locally. It wasn't until the 1920s, when farmers established irrigated fields in California's Salinas Valley, that spinach became a commercial crop.

Jefferson often received seeds from Europe, so it's likely that the spinach plants he grew were Italian varieties.

Shiver to Sprout

Spinach is a cold climate plant that thrives when temperatures are between 50°F and 60°F. Spinach can thrive in hotter weather if supplied with enough water. Slow-growing varieties planted in summer survive heat better than the fast-growing types planted in fall and winter. Spinach grows by leaps and bounds in China, which produces the most spinach in the world. Chinese farmers cultivate about 5,950,000 tons each year. California is the spinach hot spot in the United States, producing about 263,500,000 pounds per year.

Workers rest on bins during a manual spinach harvest.

Bulk spinach is harvested mechanically.

To grow what is called banded fresh spinach—the kind held together by giant twist ties—farmers use mechanical seed drills to drop about ten seeds per foot into fine, well-cultivated soil. This produces about six to eight plants per foot. After about six weeks, the first plants are ready for harvest. Farmworkers cut spinach by hand with knives, gather the leaves into bunches, and bind them with a wide twist tie.

For bulk or bagged spinach, growers tend to spread seed in a method known as broadcast. Instead of planting the seeds individually in tidy rows, farmers spread the seeds across large open areas where mechanical harvesters can easily maneuver. These machines operate much like large lawn mowers, cutting a wide area with a saw.

Banded and broadcast spinach are cut twice—once at a level of about six inches above the ground and again close to the **crown,** the dense pack of leaves just beneath the ground. Sometimes workers sort and pack spinach right in the field. If not, workers pick the crop, toss it into bins, and then rush it to the nearby packing shed for cooling, cleaning, and packing. Growers must rapidly cool spinach, like other perishable produce, to about 32°F. Growers use Hydro-Vac systems, such as those used to cool artichokes, to cool spinach.

The Raw and the Cooked

Cooked spinach is the main ingredient in exotic dishes from around the world. In India, diners eat spinach in a cheese and greens curry dish. Iranian cooks flavor spinach with cinnamon and add the green to meat stews. And don't forget about *spanakopita,* the Greek spinach and cheese pie. The pie's flaky pastry encases

To Your Health!

Spinach is packed with iron—an element essential for healthy blood. Spinach also contains beta-carotene, a powerful antioxidant that builds the immune system to fight cancer growth in the body. Beta-carotene also helps prevent strokes. Rich in folic acid, a B vitamin, spinach can be extremely important for pregnant women. Infants born to women who don't have enough folic acid can suffer from growth problems.

Spinach is used as one of many ingredients in a variety of dishes, but it is often served alone. Spinach Catalan, *(left)*, a Mediterranean dish, is flavored with raisins and pine nuts.

layers of feta cheese and spinach flavored with herbs and green onions. The Greeks also enjoy sautéed spinach topped with a cold garlic sauce. North Americans eat fresh spinach salad made with the youngest leaves of the plant, hard-boiled egg, and bits of bacon. They also sauté spinach lightly with garlic and olive oil. The French stuff mushroom caps with a spinach and ham mixture and bake them. To make *poisson à la florentine*, French cooks stuff a whole fish, such as red snapper, with spinach, cream, and cooked shallots. *Mchicha na nazi* is a Tanzanian dish of spinach, coconut milk, onion, and curry powder. Using fresh crab, coconut milk, and finely chopped spinach, cooks from Trinidad and Tobago whip up a spinach soup called *callaloo*. West Africans relish their *palava* sauce, a stew made with spinach, pumpkin seeds, fish, and meat. Cooks in the southern United States often boil greens with bits of ham or bacon. Others choose to sauté greens in olive oil and garlic.

"Greens" refer to the edible leaves of some plants, including turnips, beets, mustard, dandelions, collards, chard, kale, and spinach.

Dig In!

MCHICHA NA NAZI (SPINACH WITH COCONUT)
(4 SERVINGS)

This African dish is from Tanzania, East Africa, where diners like their spinach well seasoned.

1 large bunch fresh spinach
1 cup coconut milk—you can find this at Asian markets or most grocery stores
4 tablespoons of butter
1 medium onion, diced
1 tomato, diced
1 teaspoon curry powder

Cut off the stalks from the spinach. Rinse leaves thoroughly in cold water and pat them dry with paper towels or a clean dish towel. Pour the coconut milk into a pan. Add the spinach leaves and cook no more than two minutes on medium to low heat. Use a slotted spoon to remove the spinach to a bowl. Set aside the pan of coconut milk. In a large, shallow skillet, melt the butter over medium heat. Sauté the onion, tomato, and curry powder in the butter for about five minutes. Add the coconut milk and cook for another five minutes. Stir in the cooked spinach and cook for one more minute. Serve hot with rice and fried fish dishes.

Belgian Endive
[*Cichorium intybus*]

The word *endive* is used to describe two different plants. To the English and the Germans, endive is the plant that produces the sharp-tasting, serrated leaves that are used in salads. What is endive to the English and the Germans is chicory to the French, Belgians, and Americans. In France, Belgium, the United States, and this book, endive is the cone-shaped vegetable with crisp, tightly packed white or light yellow leaves that grow around a tiny core in the head's center. This variety of endive resembles a small ear of corn. More frequently it is called Belgian endive because the Belgians have been the plant's major developers and promoters.

Belgian endive is available in white or red.

But why don't we stop using endive only in salads? If we would stuff, braise and sauce it in the Belgian fashion, we would consume far more of it than we do.

—Michael and Frances Field

New Leaf on the Block

Appearing on the vegetable scene in the 1850s, Belgian endive is a relatively new plant. This is good news for those who wish to study it. Experimenters recorded in writing each step of endive's development. In the mid-1800s, the Dutch, Flemish, and Germans used dried chicory root, a descendant of wild, roadside chicory, as a substitute for coffee beans. Growers were constantly working to develop a bigger, better chicory root to produce more coffee-substitute from each plant.

The chief gardener at the Botanical Garden in Brussels, Belgium, was no exception. He took some chicory plants and grew them indoors in a cellar. Either deliberately or by chance, the chief gardener had built up a

You Say Witloof. . . .

Belgium has two official languages. Those who live in the north speak Flemish, a variant of Dutch, and southerners speak French. Belgian endive is called *witloof* in Flemish and *chicon* in French. Flemish gardeners developed the variety of endive most commonly grown throughout the world. It is called witloof, meaning "white leaf."

The Botanical Garden in the heart of Brussels, Belgium, is the birthplace of the Belgian endive of modern times.

Edible endive is dug up before the flower blooms. However, the plant also grows wild and produces a lovely blue flower.

mound of soil around the base of one plant. An assistant checking up on the plant was surprised to discover under the soil a long, white-leafed head of greenery—what the Belgians call "white gold." Thus was born, almost completely by accident, the crisp, tart Belgian endive of modern times. Central European farmers had planted lettucelike vegetables in dark cellars before. Doctors had recommended the leaves as a cure for kidney problems. But Belgian endive was much tastier.

When Belgian farmers began growing endive commercially, they placed the plants outside in the fields between rows of other crops. Before the endive bloomed, the growers would dig them up, roots and all. They would lop off the outer leaves, feed them to their animals, and then replant the endive roots in eighteen-inch-deep trenches in storage cellars. Farmers covered the replanted roots with soil and harvested grain husks to protect the roots from frost. After about six weeks, the farmers unearthed *witloof,* or "white leaf" in Flemish.

Before long, the Dutch began growing Belgian endive. By 1873 the French eagerly cultivated what they called *chicon.* For years many European farmers grew the plant— sometimes in large washtubs—for their own use. Commercial production took off in northern Europe in the 1930s.

Twice as Nice

Although U.S. farmers call the plant Belgian endive, northern France actually grows the most Belgian endive in the world. Chile exports some endive to the United States.

Traditionally, farmers in northern, cool-climate countries plant endive seeds in carefully tilled soil in the spring. They space the seeds four to six inches apart in rows about two feet apart. By August, plants have been growing for about 120 days and are watered by northern Europe's plentiful rains. The endives have established stocky roots and abundant leaves.

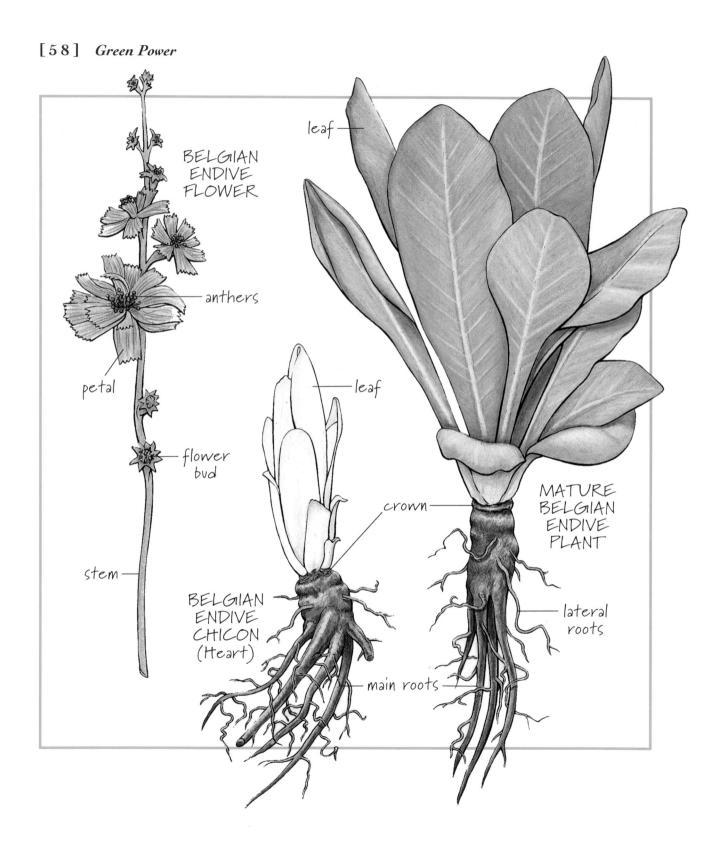

BELGIAN
ENDIVE
FLOWER

anthers

petal

flower
bud

stem

BELGIAN
ENDIVE
CHICON
(Heart)

leaf

leaf

crown

MATURE
BELGIAN
ENDIVE
PLANT

lateral
roots

main roots

Growers chop off the leaves close to the root and pull up the plants—either by hand or with small row diggers. Typically, smaller growers work by hand and larger producers use row diggers. If farmers mechanically harvest the plants, they allow the roots to dry in the field for three or four days.

To produce the whitened endive that northern Europeans have long enjoyed, Belgian growers usually replant the roots under about two-and-a-half feet of soil and cover the mounds with plastic or straw. Because the plant receives little or no light, photosynthesis—which causes the stems or leaves to turn green—cannot take place. The process of growing vegetables—such as white asparagus, pale yellow celery, and white Belgian endive—in this manner is called blanching. Those who favor blanched vegetables believe they are sweeter than anything green.

A row digger is a machine small enough to move between crop rows. It is used to pull plants from the soil.

Sometimes endive is grown **hydroponically.** To do this, farmers blanch the chicory plants by growing them in the dark, stacked in water-filled wooden bins.

In this endive field, growers use terra-cotta pots to cover endive plants. The pots create the warm, dark environment endive plants need to thrive.

Since Belgian endive aren't accustomed to growing in the dark, farmers have to **force** the plants to grow. Forcing a plant means to make a plant bloom or produce leaves earlier or more often than it would in its natural cycle. Farmers do this by creating light and temperature conditions that extend or speed up a plant's growing cycle.

In the case of Belgian endive, growers must create favorable conditions to promote growth. Without natural sunlight, for example, the plants need steady warmth to grow. So farmers bury water pipes beneath the roots and pump hot water through them to maintain a constant temperature of about 68°F. After only three or four weeks, workers dig out the plants and break the leaves from the roots. They trim around the base of the endive plant to remove any loose leaves. Finally, they wash the endive in centrifugal drums. The drums' tumbling motion helps knock the dirt off the endive.

Since the 1960s, some farmers have grown endive using hydroponics, that is, with water instead of soil. Growers first establish the endive roots in trays filled with gravel. They set the trays in dark forcing rooms and feed the plants water and fertilizer. Growers keep

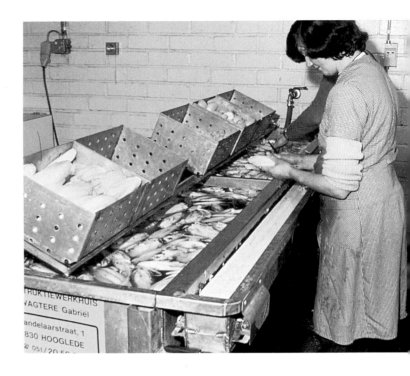

A worker in Belgium soaks and cleans harvested endive.

Growers often feed endive roots to their animals.

the room at a temperature of between 55°F and 65°F. To allow for easy harvest, farmers keep the trays stacked in frames. Farmers cut the mature, torpedo-shaped leaves from the roots. Nothing goes to waste.

Hydroponically grown endive matures at the same rate as plants grown in soil. One advantage to growing endive hydroponically is that farmers don't need to apply **pesticides.** Pests that live in soil aren't able to survive in water. But some experts feel that this method reduces flavor. Critics believe that only the "forced under the soil" method produces that distinctive sharp yet sweet taste endive lovers want.

To Your Health!

Belgian endive is a good source of fiber and is packed with potassium, which regulates the body's balance of water. Endive's most valuable contribution, however, is selenium. Endive contains a small amount of selenium, which is derived from healthy soil. In these small amounts, selenium improves oxygen levels in the blood. The endive's roots store selenium, then release it into the leaves during the forcing process.

Endive is a fresh and tangy addition to any salad.

At processing plants, workers sort and grade all endive according to size, tightness of leaves, and amount of yellow or yellowish green showing on top of the leaves. Because very white is most desirable, the whitest of the white are the most expensive. Workers wrap these leaves in layers of purple tissue paper and pack them in wooden crates. The colored paper protects the endive from discoloring in the light.

Dig In!

Braised Belgian Endive
(4 servings)

8 Belgian endive
2 tablespoons butter
3 tablespoons fresh lemon juice
¼ teaspoon salt
1 tablespoon sugar

Select eight firm Belgian endive with yellow tips. Wash them in cold water and pat them dry on a clean dishtowel or paper towel. About half an inch from the bottom of each leaf, cut off the end of the endive. Use a knife or potato peeler to remove the very bottom of the inner core.

Melt the butter in a heavy, shallow frying pan. Add the endive and cook lightly on high heat, turning them to cook each side about a minute and a half. Sprinkle the remaining ingredients over the endive and cover the pan tightly. Turn the heat down to a low simmer and cook about twenty-five minutes. Serve hot.

Try It, You'll Like It

Belgians braise endive by cooking it lightly in butter, sprinkling it with lemon juice, and then covering and cooking it on low heat until tender. Belgians also enjoy endive chopped and mixed with leeks and potatoes in a creamy soup. Or they might braise the vegetable, stuff it with chicken, wrap it in ham, and then bake it in a rich white sauce made from butter, flour, and milk. In the summer, endive tastes good in a fresh salad, sliced thinly with a hard-boiled egg and served with vinaigrette dressing.

Lettuce

[Lactuca sativa]

Lettuce, the ultimate leafy green, has many varieties. A typical North American seed catalog offers eight different types of leaf lettuce, for example. There are three main varieties of lettuce—head, leaf, and cos. Head lettuce is probably the most common type of lettuce found on tables throughout North America. Crisp head lettuces such as iceberg form tight, layered, well-rounded heads that look a bit like cabbages. Butterhead varieties, such as Boston lettuce, are also considered head lettuces even though their soft leaves form extremely airy heads. Leaf lettuces flop in the breeze and don't form firm heads. This type is easiest to grow and is the lettuce you're likely to find in home gardens. Lettuce plants that grow upright, with tall, slender leaves are known as cos lettuces. Romaine is the most popular type of cos lettuce.

Red and green leaf lettuces

Lettuce is like conversation—it must be fresh and crisp, and so sparkling that you scarcely notice the bitter in it.

—Charles Dudley Warner

Lettuce Begin

Lettuce is first mentioned in an **herbal**—a book about medicinal plants—written in Babylon in 800 B.C. A plant happiest in cool temperatures, lettuce more than likely did not originally come from the desert regions of Babylon. Experts have guessed that lettuce may have originated in Siberia, the Mediterranean region, or the Middle East.

From the writings of Herodotus, a Greek historian who lived between 484 and 432 B.C., we know that the Persians ate lettuce. Egyptians ate the plants, too. Archaeologists studying the ancient Egyptians have found lettuce seeds in Egyptian tombs. In ancient Egypt, seeds symbolized hope for the rebirth and renewal of the dead person.

The lettuce the ancient Egyptians grew was nothing like what we eat in modern times. Egyptians ate the small leaves of the plant's tall seed stalks. If you've ever planted lettuce at home and seen it **bolt,** or send up seed stalks once hot weather arrives, you've seen ancient Egyptian lettuce.

The Greeks ate lettuce at this stage, too. But they called it "asparagus," a word they used for many stalklike vegetables. It wasn't until at least A.D. 300 that the Romans tamed the stalks into heads. The Romans, who ate the leaves with salt, gave us the concept and the word *salad*, which comes from *herba salata* or "salted greens." Many Romans ate lettuce with olive oil dressing, sometimes to start a meal and other times to close the meal.

The ancient Egyptians ate lettuce with their meals. This drawing on the wall of a tomb in Thebes, Egypt, portrays a servant hunting for food.

The ruins of Diocletian's estate still surround the town of Split, Croatia.

To Your Health!

Lettuces are high in fiber and rich in beta-carotene, vitamin C, and vitamin E. Vitamins C and E help open arteries to allow for healthy blood flow. Beta-carotene boosts the immune system to fight cancer. Lettuces also contain phytochemicals— nutrients that provide more protection against cancer.

Diocletian, Roman emperor from A.D. 284 to 305 and best known for persecuting Christians, was a gardener and a lettuce lover. He quit his job as emperor and retired to what later became the town of Split, Croatia, to grow lettuce. Maximian, his successor, feared for his own life and begged Diocletian to return to office. Diocletian responded, "If you saw what beautiful lettuces I am raising, you would not urge me to take up that burden again."

Much of the lettuce that Roman gardeners cultivated bolted and went wild after A.D. 476, when the Roman Empire fell. In later centuries, monks began to cultivate the plant once more, and others simply picked the wild leaves. The Chinese grew lettuce beginning in the A.D. 400s, but we don't know much about its travels elsewhere in Asia.

During the Middle Ages, which began in Europe around A.D. 500, people grew lettuce in small home gardens. They munched on crisp lettuce salads as often as we enjoy them in modern times. By the 1000s, towns had sprung up along rivers, and trade between towns flourished. During this time, the Roman Catholic Church played a central role in people's lives. Priests traveling from one Roman Catholic outpost to the next carried many plants and seeds with them.

In the 1300s, Italian monks brought a new, tall lettuce from Rome to France. The French called this tall variety of crisp lettuce Romaine, because it came from Rome. The Romans themselves called it cos lettuce, after the sunny Greek island of Cos, where the lettuce first grew. The English didn't begin cultivating lettuce until the 1400s, and they didn't sell the vegetable in markets until late in the 1500s.

Cesar Chavez

From the beginnings of commercial lettuce production in California, farmworkers have fought for fair wages and safe working conditions. Workers asked for fair working hours with breaks, plentiful drinking water, safe sanitation, protection from chemicals and pesticides, and clean, safe housing for migrant workers—workers who came north to work in the fields for the growing season and then went back to Mexico.

The beginning of World War II (1939–1945) brought a shortage of male workers. At the same time, American workers were beginning to join unions—groups dedicated to improving working conditions—and to demand higher pay. The U.S. government made it possible for Mexican workers to enter the country legally to take picking and packing jobs in California and elsewhere. Employers were able to pay the Mexican workers less than American workers. By 1959, 72 percent of the seasonal laborers in Monterey County, California, were Mexican.

In the early 1960s, labor leader Cesar Chavez headed the United Farm Workers union. Chavez was the child of Mexican farmworkers. He grew up near Yuma, Arizona, in camps for migrant laborers. In 1970 Chavez worked to draw attention to the tough working conditions of farmworkers. He staged a national boycott—an organized refusal to buy an item—of lettuce grown by nonunion workers. But the boycott didn't work. Workers were often afraid to unionize. Chavez went on hunger strikes—a form of protest in which participants refuse to eat until the group they are protesting against follows through with protester demands. Chavez even spent time in jail to draw national attention to the poor working conditions. Chavez died in 1993. In 1994 President Bill Clinton awarded Chavez the Presidential Medal of Freedom to honor his fight for social justice and commitment to dignity and fairness for farmworkers.

THREE MAIN
LETTUCE VARIETIES

leaf

head

HEAD
LETTUCE

LEAF LETTUCE
(a red variety)

COS
LETTUCE

roots

During the harvest, workers walk behind a tractor and pack the lettuce into boxes. Growers must transport the boxes right away as lettuce wilts very rapidly.

No one is sure who brought the first lettuce seeds to North America. Christopher Columbus may have carried some with him when he arrived in the Caribbean in 1492. Or Spanish colonists may have brought some along in the late 1500s. Because seeds for hundreds of lettuces fit into a small paper packet, any number of people probably carried lettuce seeds west. The English, Dutch, and other immigrants to North America certainly had lettuce in the early 1600s, as the plant was essential to any settler's garden.

At the same time, salads became exotic affairs in Europe. Europe's royalty became lettuce-mad. King Louis XIV of France munched his lettuce seasoned with basil and tarragon. The English aristocrats of the following century loaded lettuce with fish preserved in oil and topped with mustard. Among French royalty in the 1600s, salmagundi—a sculptural salad built layer by layer rather than tossed together—became popular. Greens, chopped meats and vegetables, pickles, and hard-boiled eggs make up a salmagundi.

On the other side of the Atlantic, North American cooks were developing their own leafy concoctions. Waiters at the Waldorf-Astoria Hotel in New York served up the first Waldorf Salad in 1893. In 1924 an Italian chef at Caesar's Restaurant in Tijuana, Mexico, tossed together what came to be called a Caesar salad. Robert Cobb, owner of

Tunisian families gather at the market. Tunisians use lettuce in different types of salads.

the Brown Derby Restaurant in Hollywood, California, padded into the kitchen late one night in 1934 for a snack. Chopping up veggies he had on-hand, Cobb created what would later be dubbed the Cobb salad.

A Seedy Bunch

Lettuce is cultivated all over the world, but China produces the most lettuce by far. Chinese producers crank out an average of 6,065,000 tons of the tasty leaves per year. The United States ranks second, producing about 4,355,000 tons per year. Most of the lettuce is grown in California.

California's full-scale commercial operations took off in the 1930s. California supplies two-thirds of the lettuce in the United States. In modern times, farmers and gardeners grow lettuce everywhere in North America. On many ten-thousand-acre farms in California's Salinas Valley, lettuce thrives virtually year-round with the help of irrigation. Commercial growers in Arizona also raise lettuce twelve months of the year. Growing lettuce is labor intensive. Each spring, workers plant lettuce seeds by hand twelve inches apart in rows from thirty to forty-two inches wide. Farmworkers use a long-handled hoe to weed between the plants. After about seventy days of growth, the lettuce is ready for harvest.

Workers use a machine to transplant lettuce seedlings in a field of romaine lettuce.

To harvest the lettuce, a worker drives a tractor that pulls a boom through the rows. Six or more workers walk behind the boom and bend down to cut the heads with a wide, rounded knife. Workers will often cut fields twice, especially when harvesting bulk lettuce—the leafy varieties rather than those that grow in heads. Workers in the field might then hand up lettuces to workers on the boom who trim the outer leaves and pack the heads into cartons. Otherwise, workers place the heads on a conveyor belt that leads to the nearby packing area. Workers pack some lettuce, usually the loose-leaf variety, naked—without plastic covering. Iceberg is usually wrapped in plastic. Workers toss bulk lettuce into large bins—the industry uses plastic bins that hold one thousand pounds of lettuce.

In the United States, iceberg lettuce is grown mainly in California. Iceberg got its name in the 1930s, when it had to be packed in block ice for shipment from California to eastern markets.

(Above) A worker prepares to close a vacuum cooler, which keeps the lettuce fresh in storage. (Right) Workers harvest a greenhouse crop of various lettuces.

Growers must quickly cool all lettuce, wrapped or not. They usually use the same water and vacuum process used to chill broccoli. Workers store packed and cooled lettuce in huge cold rooms until refrigerated trucks pick up the shipment for delivery.

North American cafeterias and fast food restaurants serve more iceberg lettuce than any other type. Its tighter, cellophane-wrapped head means fewer bruises in transit and a longer shelf life. But iceberg, a light green lettuce, packs fewer nutrients than other, darker green lettuces do. Dark green leaves contain more cancer-fighting elements than lighter green lettuces do.

Crunch on This

Lettuces are best when eaten fresh and tossed in dressing, although they can be lightly cooked. The secret to a tasty salad is to throw together many different types of lettuce—try the tart with the sweet, or mix red varieties with green types.

Over the years, people who want to eat more than just leaves have spiced up salads. Oscar Tschirky, head waiter at New York's famed Waldorf-Astoria Hotel, created the first Waldorf Salad. He mixed together apples, celery, and mayonnaise and served them on a bed of iceberg lettuce. In later years, cooks added walnuts, too.

Plant and Snip

Try growing your own lettuce! It's easy. A window box works well. You can grow lettuce on an inside windowsill in winter, especially if you rotate the container every week. Lettuce is also easy to grow in a large pot outdoors in spring. Here's how:

Place a layer of pebbles on the bottom of the pot or window box to provide good drainage. Fill the pot with fresh potting soil. Use a watering can or hose to wet the soil thoroughly. Scatter the lettuce seeds—you can buy a mix especially for cutting—on top of the soil. Cover the seeds with 1/4 inch of soil. Pat the soil down gently with your hand. Keep the soil moist by spritzing daily with a spray bottle filled with water. When the lettuces peek through the soil about seven days later, water them gently with a hose or watering can. Once the leaves are about four inches tall, use a scissors to cut the amount you want. Allow the leaves to grow back and cut them again. You can do this about three times before the lettuce stops producing tasty leaves.

Alex Cardini, an Italian chef, created the original Caesar salad in 1924 at his brother Caesar's Restaurant in Tijuana, Mexico. Cardini supposedly threw together the salad—Romaine lettuce, vinaigrette dressing, Parmesan cheese, croutons, and egg—one Fourth of July for a group of people from California when he could find little else in the kitchen. Eventually he named the creation for his brother. Later on, other chefs added anchovies and Worcestershire sauce.

The Cobb salad is a mix of lettuce, bacon, and avocado, tossed together in a bowl with a Worcestershire vinaigrette. Robert Cobb later refined the salad. He added tomatoes, chicken, egg, chives, and blue cheese, and a classic dish was born. Cobb salad mixes bibb, Romaine, chicory, and watercress greens.

People's preferences for salad dressing are as varied as their tastes for different dishes. The Greeks favor a rich oil, vinegar, and herb dressing mixed with feta, a crumbled goat's cheese. In Tunisia and Morocco, dressings often include lemon juice and sugar. Cooks in Thailand and Indonesia like to flavor dressings with limes and hot chilies. Lettuce salads are not traditional in Japan, but they are becoming more popular. Japanese dressing might include fresh ginger, soy sauce, sesame oil, and rice vinegar. The Finns blend yogurt, fresh dill, cucumbers, vinegar, oil, and minced onion to make a refreshing dressing. French chefs make their classic vinaigrette right in the salad bowl. First they rub the bowl with a raw garlic clove. Then they add black pepper, salt, and a dab of Dijon mustard. Finally cooks add olive oil and whisk the ingredients together.

Cobb salad

Greek salad

Dig In!

WALDORF SALAD
(3 SERVINGS)

This famous salad has many variations. This basic version is easy and tasty.

1 cup diced celery (about 4 stalks)
1 cup diced apples (about 2 apples)
½ cup chopped walnuts
¾ cup mayonnaise
3 lettuce leaves

Wash apples and celery well before dicing—cutting into small pieces about ⅛ to ¼ inch thick. Toss together all the ingredients except lettuce and chill in the fridge for at least two hours. To serve, place a large, rounded spoonful of the chilled mix on each lettuce leaf.

Glossary

blanch: To block sunlight from reaching a plant, thereby preventing photosynthesis from taking place and keeping the plant pale in color.

bolt: To produce seed prematurely.

botanist: A scientist who specializes in the study of plants.

cover crop: A crop planted in between seasons to prevent soil erosion and to enrich the soil.

crown: A first-year growth of roots and underground stems that can be planted instead of seeds to grow some perennial crops.

domestication: Taming animals or adapting plants so they can safely live with or be eaten by humans.

force: To make a plant or fruit grow faster than is natural by giving it extra heat or light.

hand-pollinate: Placing pollen on a flower by hand so that the flower will grow into a fruit or vegetable.

herbal: A book, often illustrated, about plants that typically discusses the medicinal value of each plant.

hybrid: The offspring of a pair of plants or animals of different varieties, species, or genera.

hydroponics: Growing plants in water and minerals instead of in soil.

ornamental: A plant grown for its beauty and not for its food or commercial value.

perennial: A plant that lives for more than two years.

pesticides: Poisons that growers apply to crops in order to kill unwanted insects. Pesticides can also harm humans and animals.

photosynthesis: The chemical process by which green plants make energy-producing carbohydrates. The process involves the reaction of sunlight to carbon dioxide, water, and nutrients within plant tissues.

Further Reading

Brennan, Georgeanne, and Ethel Brennan. *The Children's Kitchen Garden: A Book of Gardening, Cooking, and Learning.* Berkeley, CA: Ten Speed Press, 1997.

Fitzsimons, Cecilia. *Vegetables & Herbs.* New York: Julian Messner, 1997.

Kite, L. Patricia. *Gardening Wizardry for Kids.* Hauppauge, NY: Baron's, 1995.

Nottridge, Rhoda. *Vitamins.* Minneapolis: Carolrhoda Books, Inc., 1993.

Root, Waverley. *Food.* New York: Simon and Schuster, 1980.

Tames, Richard. *Food: Feasts, Cooks, and Kitchens.* New York: Franklin Watts, 1994.

Trager, James. *The Food Chronology.* New York: Henry Holt and Company, 1995.

Wake, Susan. *Vegetables.* Minneapolis: Carolrhoda Books, Inc., 1990.

An Indonesian woman includes greens among her wares.

Index

About the Author

Meredith Sayles Hughes has been writing about food since the mid-1970s, when she and her husband, Tom Hughes, founded The Potato Museum in Brussels, Belgium. She has worked on two major exhibitions about food, one for the Smithsonian and one for the National Museum of Science and Technology in Ottawa, Ontario. Author of several articles on food history, Meredith has collaborated with Tom Hughes on a range of programs, lectures, workshops, and teacher-training sessions, as well as *The Great Potato Book.* The Hugheses do exhibits and programs as The FOOD Museum in Albuquerque, New Mexico, where they live with their son, Gulliver.

Acknowledgments

For photographs and artwork: © Steve Brosnahan, p. 5; Tennessee State Museum Collection, detail from a painting by Carlyle Urello p. 7; Alinari/Art Resource, NY p. 12; © Trip/Streano/Havens, p. 15 (top); © Bettman/CORBIS, pp. 15 (bottom), 68; © Inga Spence/TOM STACK AND ASSOCIATES, pp. 17, 29; © D. Cavagnaro/Visuals Unlimited, p. 18; © Joe Rychetnik/The National Audubon Society/Photo Researchers, p. 19; © Trip/H. Rogers, pp. 20 (top), 42, 47 (bottom); © Jacques M. Chenet/CORBIS, p. 20 (bottom); © Walter/Louiseann Pietrowicz/September 8th Stock, pp. 21, 31, 43, 53, 63, 75 (both), 76; ©Voscar/AGStock USA, p. 23; © Charles & Josette Lenars/CORBIS, p. 25; © Craig Lovell/CORBIS, p. 26; © Morton Beebe, S.F./CORBIS, p. 27 (bottom); © Ed Young/AGStock, p. 27 (top); © Trip/R. Nichols, p. 30; © Mark E. Gibson/ Visuals Unlimited, p.33; © Archivo Iconografico, S.A./CORBIS, pp. 34, 35 © Joseph Sohm; ChromoSoh Inc./CORBIS p. 36; © Stephanie Maze/CORBIS, p. 38 (left); © Bill Kamin/Visuals Unlimited, p. 38 (right); © Jerry Pavia, p. 39 (top and bottom); © Michael S. Yamashita/CORBIS, p. 40; © Super Stock, Inc., pp. 41, 52; © Dwight R. Kuhn, pp. 45, 65; © British Museum/The Art Archive, p.46; © Patricia Ruben Miller/IPS, p. 47 © Roger Bennett/AGStock USA, pp.49, 50 (bottom); © Mark S. Skalny/Visuals Unlimited, p. 50 (top); © G. Buttner/Naturbild/OKAPIA/The National Audubon Society/Photo Researchers, p. 55; © Gillian Darley; Edifice/CO/CORBIS, p. 56; © The Purcell Team/CORBIS, p. 57 © Holt Studios/Richard Anthony, pp. 59, 73 (top and bottom); © Holt Studios/Stuart Lumb, p. 60; © Belgian National Tourist Office, New York, p. 61; © Michelle Garrett/CORBIS p. 62; © Gianni Dagli Orti/CORBIS, p. 66; © Tony Hertz/AGStock, p. 70; © Trip/B. Turner, p. 71; © Macduff Everton/CORBIS, p. 72; © Trip/J. Perry, p. 78. All other artwork by Laura Westlund. Cover Photo by Jim Simondet.
For quoted material: p. 4, M.F.K. Fisher, *The Art of Eating* (New York: Macmillan Reference, 1990); p. 10, Edward Lear, "The New Vestments," as quoted by Brigid Allen, ed. *FOOD* (Oxford, U.K.: Oxford University Press, 1994); p. 22, Roy Blount, Jr., "Against Broccoli," as quoted in *The Penguin Dictionary of Modern Humorous Quotations* (New York: Penguin Books, 1987); p. 34, *Historia Naturalis* (77 A.D.); p. 32, Ogden Nash, *FOOD* (New York: Stewart, Tabori and Chang, 1989); p. 54, Michael and Frances Field, *A Quintet of Cuisines* (New York: Time-Life Books, 1970); p. 64, Charles Dudley Warner, "My Summer in a Garden," as quoted by March Egerton, *Since Eve Ate Apples* (Portland, OR: Tsunami Press, 1994); p. 67, Waverley Root, *Food* (New York: Simon & Schuster, 1980).
For recipes: (some slightly adapted for kids): pp. 21, 31, 43, 63, 76; Meredith Sayles Hughes; Dorinda Hafner, *A Taste of Africa* (Berkley, CA: Ten Speed Press, 1993).